Animals Live Here

Life in the Mountains

By Connor Stratton

www.littlebluehousebooks.com

Copyright © 2020 by Little Blue House, Mendota Heights, MN 55120. All rights reserved. No part of this book may be reproduced or utilized in any form or by any means without written permission from the publisher.

Little Blue House is distributed by North Star Editions:
sales@northstareditions.com | 888-417-0195

Produced for Little Blue House by Red Line Editorial.

Photographs ©: D'July/Shutterstock Images, cover; ovbelov/Shutterstock Images, 4; AndreAnita/iStockphoto, 7 (top), 24 (top right); Kelp Grizzly Photography/Shutterstock Images, 7 (bottom); Astrid Gast/Shutterstock Images, 9; Andranik Barsegyan/iStockphoto, 11; tahir abbas/iStockphoto, 12; Bob Cullinan/Shutterstock Images, 15 (top); Ondrej Prosicky/Shutterstock Images, 15 (bottom); iliuta goean/Shutterstock Images, 16–17, 24 (bottom left); carlosobriganti/Shutterstock Images, 18; Dennis W Donohue/Shutterstock Images, 21 (top), 24 (bottom right); Harald Toepfer/Shutterstock Images, 21 (bottom); Jennifer Placek/Shutterstock Images, 23, 24 (top left)

Library of Congress Control Number: 2019908611

ISBN
978-1-64619-022-5 (hardcover)
978-1-64619-061-4 (paperback)
978-1-64619-100-0 (ebook pdf)
978-1-64619-139-0 (hosted ebook)

Printed in the United States of America
Mankato, MN
012020

About the Author

Connor Stratton enjoys hiking up mountains, spotting new animals, and writing books for children. He lives in Minnesota.

Table of Contents

Mountain Animals **5**

Birds **13**

Mammals **19**

Glossary **24**

Index **24**

Mountain Animals

The mountains are high up and cold.

Yaks live here.

Many other animals live here too.

Goats live in the mountains. They have horns and hooves. They can climb up mountains and jump across rocks.

Marmots live in the mountains.

They have big front teeth.

Butterflies live in the mountains. They fly with their colorful wings.

Birds

Birds live in the mountains. Some can fly as high as a mountain.

Crows live in the mountains. These birds have black feathers. Vultures live in the mountains too. These birds have long necks and large wings.

Eagles live in the mountains. These birds have sharp claws. They use the claws to grab small animals.

Mammals

Many mammals live in the mountains.

Mammals have fur or hair.

Leopards live in the mountains. These mammals have spotted fur. Llamas live in the mountains too. These mammals have long necks.

Bears live in the mountains. These mammals have thick fur and sharp teeth. They can roar loudly.

Glossary

bear

goat

eagle

leopard

Index

L
llamas, 20

M
marmots, 8

V
vultures, 14

Y
yaks, 5